DEMYSTIFYING
AI
for Business
Executives

Henry Collins
Xavier Taylor II

Book Author: Henry Collins
Book Designer : Peyton Winder
Art Director / Coauthor : Xavier Taylor II
Cover Art: Xavier Taylor II
Interior images promoted by : Peyton Winder / Henry Collins

Contents

Dedication

To Jehovah God, the true source of AI (absolute intelligence).

To my newborn son Xander, you are my perpetual happiness.

Acknowledgments

Thank you, Melissa, for being patient during this project.

I want to acknowledge my wonderful wife, Katrina, for always believing in the creative process.

Foreword

In today's ever-evolving landscape of Artificial Intelligence (AI), writing a current and up-to-date book is like filling an Olympic-size pool with a 2-quart pitcher. This field grows and transforms so rapidly that any attempt to document its current state risks obsolescence the moment it is published. Yet, this does not diminish the value of the endeavor; rather, it highlights the critical need for a foundational understanding of AI in the business world today.

The intention behind this book is not to serve as an exhaustive compendium of AI's capabilities and technical intricacies. Instead, it is designed to be your guide through the complex terrain of AI technology, offering a clear overview and introduction to the subject. It aims to demystify AI for business executives like yourself, providing the essential knowledge needed to navigate this dynamic field confidently.

Why is there a vital need to incorporate AI into your daily workflow? The answer is twofold. Firstly, AI is no longer a futuristic concept but a present reality that is reshaping industries, streamlining operations, and enhancing decision-making processes. Its influence permeates every sector, making an understanding of its mechanisms and potential applications indispensable for any business leader. Secondly, the agility offered by AI technologies allows businesses to adapt more swiftly to market changes, predict consumer behavior with greater accuracy, and innovate at unprecedented speeds. This book is designed to be your compass in the AI landscape, helping you understand its language, recognize its potential, and foresee its impact on your business strategies. It seeks to arm you with the knowledge to make informed decisions about integrating AI technologies into your operations, ensuring that your business remains competitive in a rapidly changing world.

As you embark on this journey through the pages of "Demystifying AI for Business Executives," keep in mind that the essence of AI's value to business executives lies not in the minute details of its algorithms or the latest advancements in machine learning techniques. Rather, it is in understanding the strategic implications of AI for your business, the ethical considerations it entails, and the leadership required to navigate its challenges and opportunities.

We acknowledge that the field of AI will continue to evolve, perhaps in ways we cannot yet anticipate. However, the foundational knowledge in this book is designed to equip you with the ability to adapt to and embrace these changes, fostering a culture of innovation and forward-thinking within your organization. By embracing the insights shared within these pages, you position yourself and your business at the forefront of the AI revolution, ready to harness its potential for growth and success. Welcome to your AI journey. Let this book be the first step in demystifying a technology reshaping the future of business—and daily life.

Demystifying AI for Business Executives

Collins | Taylor II

Just what is Artificial Intelligence?

Topics

Just What is Artificial Intelligence

The Evolution of AI in Business

Chapter 1

Just what is
Artificial Intelligence?

○────────────────────────────○

In this chapter, we will simplify the concept of Artificial Intelligence (AI) and its implications, demystifying it for business executives. The term artificial intelligence (AI) has become a buzzword in recent years, but what does it truly mean?

The term "artificial intelligence" (AI) is commonly used to describe a machine's increasing ability to replicate human intelligence and take on tasks that were previously reserved for humans. Machine learning, natural language processing (NLP), robotics, and computer vision are just a few of the many methods and technologies under this umbrella. Thanks to these advancements, machines can now analyze data, make decisions, spot patterns, generate images from text prompts, and solve difficult issues.

Artificial intelligence (AI) is revolutionizing businesses across all sectors by facilitating better decision-making and improving process automation and customer service. Leveraging the power of AI is critical for businesses to remain relevant and competitive in today's marketplace and to drive internal innovation.

Machine learning is an essential part of artificial intelligence; it entails teaching computers to learn from data without being explicitly programmed. New insights, trends, and predictions can all be mined from a company's data using this technology.

For instance, recommendation systems driven by AI can sift through client data to make tailored product suggestions that boost revenue and satisfaction.

The ability of machines to understand and interpret

human language is made possible by natural language processing (NLP) in the form of chatbots like ChatGPT-4 and Claude. This technology is evolving rapidly and has already given birth to a plethora of bots and applications that can completely automate everyday workflows and tasks. Businesses can use bots to create virtual agents to handle customer service inquiries, evaluate social media feedback, and glean insights from unstructured data sources such as customer reviews and legal papers.

Artificial intelligence also plays a major part in robotics and automation, which boosts efficiency and output in sectors as diverse as manufacturing, logistics, and healthcare. AI-powered robots can handle laborious or hazardous tasks, allowing more time for product innovation and increased output efficiency. Nonetheless, it should be noted that AI also gives rise to ethical and societal worries. Bias in AI algorithms, data privacy issues, and potential job displacement are some of the challenges that need to be addressed. Understanding these consequences as company leaders is essential for utilizing AI for responsible and sustainable expansion.

In conclusion, artificial intelligence is a cutting-edge tool with game-changing potential for organizations of all sizes thanks to its capacity to streamline operations, enhance decision-making, and provide better customer service. To fully benefit from this revolutionary technology, intrepid business leaders must adopt AI and have a firm grasp of its strengths and weaknesses.

The Evolution of AI in Business

Artificial Intelligence (AI) is rapidly reshaping today's world in ways that are yet to be discovered. Industries and businesses are now using AI to restructure the way they operate. In this section, we will explore the evolution of AI in business, charting its growth and exploring its potential for businesses.

In its early stages, termed AI 1.0, the technology was primarily used to automate repetitive operations and increase efficiency. Tools like chatbots and virtual assistants were launched to handle client inquiries, reducing reliance on human labor. This automation not only resulted in significant cost savings but also improved the customer experience by streamlining operations.

As the technology progressed into the era of AI 2.0, its applications expanded substantially, including predictive analytics and data-driven decision-making. Using massive datasets, AI algorithms have become excellent at understanding patterns, predicting trends, and projecting events with astonishing precision. This skill has enabled organizations to predict client behavior, fine-tune supply chain operations, and tailor marketing strategies, boosting profitability and gaining a competitive advantage. Additionally, artificial intelligence has altered the field of robotics, allowing for the creation of autonomous devices capable of performing complicated tasks with precision and speed. Artificial intelligence-powered robots are increasingly being used in industries such as manufacturing, logistics, transportation, and healthcare, resulting in better productivity and safety.

The transition to AI 2.0 represented a logarithmic leap forward in the evolution of AI technology. This phase is defined by the emergence of autonomous Artificial General Intelligence (AGI) systems that are capable of learning, understanding, and functioning across a broad range of human cognitive abilities. Unlike their predecessors, these AGI systems are not confined to narrow, specific tasks but can adapt and apply their intelligence to various challenges, mirroring human-like versatility. AI 2.0 technologies are seamlessly being integrated into everyday life, revolutionizing how we interact with our environment, enhancing decision-making processes, and enabling a future where human and machine intelligence coexist in a symbiotic relationship. The advent of AI 2.0 marks a significant milestone, heralding a new era of technological innovation and societal transformation. Looking ahead, the future of AI in business holds immense potential. As AI progresses, its capabilities will expand, further enabling businesses to unlock new opportunities and tackle complex challenges. From autonomous vehicles and smart cities to personalized medicine and intelligent virtual assistants, AI will play a pivotal role in shaping the future of business.

To summarize, the growth of AI in business has been spectacular, altering industries and revolutionizing how businesses work. AI has become an indispensable tool for corporate executives, capable of automating operations, anticipating customer behavior, and enabling autonomous systems. Executives that embrace AI technology and leverage its potential can open up new paths for growth, innovation, and success in the ever-changing corporate world.

Chapter 2

Topics

Natural Language Processing in Business Operations

Multi Model AI and its Application in Business

Chapter 2

Machine Learning and its Applications in Business

In recent years, the field of artificial intelligence (AI) has witnessed remarkable advancements, with machine learning emerging as a powerful tool that has revolutionized various industries. As business executives, understanding the applications and potential of machine learning is crucial in staying competitive and harnessing its benefits effectively.

Machine learning can be defined as a subset of AI that enables computer systems to learn and improve from data without being explicitly programmed. This technology can analyze vast amounts of data, identify patterns, and make accurate predictions or decisions based on that information.

It has already found numerous applications across various business sectors.

One of the most prevalent applications of machine learning in business is customer relationship management (CRM). By analyzing customer data, machine learning algorithms can predict customer behavior, shopping preferences, and potential churn. This information allows businesses to tailor their marketing strategies, personalize customer experiences, and improve customer retention rates.

Another area where machine learning is making a significant impact is in fraud detection and prevention. Machine learning algorithms can analyze patterns and detect anomalies in large datasets, helping businesses identify potentially fraudulent activities and take necessary actions to mitigate risks. This is particularly relevant for financial institutions, e-commerce platforms, and healthcare providers, where the cost of fraud can be substantial.

Supply chain optimization is another domain where machine learning is proving invaluable. Machine learning algorithms can analyze historical data, forecast demand, optimize inventory levels, and enhance logistics operations. This not only improves operational efficiency but also reduces costs and improves customer satisfaction.

24   Chapter 2: Machine Learning and its Applications in Business

Machine learning is also transforming the field of medicine. For example, medical researchers at Health Data Research UK (HDR UK) are developing AI systems that can accurately detect diseases like cancer, neurological disorders, and heart conditions from medical imaging. These systems frequently spot subtle patterns in the images that human eyes might miss, leading to earlier and more precise diagnoses.

In conclusion, machine learning has immense potential to drive innovation and improve business operations across various sectors. As business executives, understanding the applications and benefits of machine learning is crucial in harnessing its power to gain a competitive edge. By leveraging machine learning, businesses can enhance customer experiences, optimize operations, mitigate risks, and drive overall growth and profitability.

Natural Language Processing (NLP) in Business Operations

One area where AI is making significant strides is Natural Language Processing (NLP). NLP is a branch of AI that focuses on the interaction between computers and human language, enabling machines to understand, interpret, and generate human language.

For business executives, understanding NLP and its applications is crucial for harnessing the full potential of AI technology. NLP can revolutionize various aspects of business operations, from customer service and marketing to data analysis and decision-making.

One of the key applications of NLP in business is customer service. With NLP-powered chatbots and virtual assistants, organizations can provide round-the-clock support to their customers without using off-shore talent with limited native language skills. These intelligent systems can understand and respond to customer queries, resolve issues, and even personalize interactions. By automating customer service, companies can enhance customer satisfaction, reduce costs, and improve overall efficiency.

NLP can assist in data analysis and decision-making processes. By using advanced text analytics techniques, organizations can extract pertinent information from unstructured data sources such as emails, reports, and contracts. NLP algorithms can categorize and summarize this information, enabling executives to identify patterns, trends, and anomalies quickly. This provides decision-makers with actionable insights, helping them make informed choices and promote business growth.

To make the most of NLP, business executives

need to understand its limitations and challenges. Language nuances, cultural differences, and ambiguous expressions can confuse NLP algorithms. To guarantee accurate and trustworthy results, it is crucial to collaborate closely with data scientists and AI specialists to address these possible issues.

In conclusion, understanding and leveraging the power of NLP will enable you to unlock the true potential of AI technology and gain a competitive advantage in your market sector. However, early adoption and innovative use cases will be a major factor for success.

Multimodal AI and its Implications for Business

Recent advancements in the field of artificial intelligence have been particularly notable in the development of multimodal AI. This innovative technology enables machines and computer systems to process and understand a combination of text, audio, and visual data, mirroring human abilities to interpret and analyze diverse forms of information. Integrating these varied data types allows multimodal AI to perform complex analytical operations and tasks, marking a significant leap forward in AI capabilities.

Multimodal AI's ability to seamlessly process and

synthesize information from text, audio, and visuals opens unprecedented possibilities across various industries, significantly impacting business strategies and operations. For business leaders, embracing the potential of multimodal AI is critical for staying ahead in today's rapidly evolving, data-driven marketplace. The technology's applications span numerous sectors, including but not limited to retail, healthcare, manufacturing, and security.

Retail

In the retail sector, multimodal AI can revolutionize customer experience by providing personalized recommendations that consider textual feedback, auditory cues, and visual preferences. For instance, a fashion retailer could utilize multimodal AI to analyze customer reviews (text), spoken feedback (audio), and images of preferred styles (visual) to offer highly customized clothing suggestions. Additionally, multimodal AI can enhance inventory management through the automated tracking and analysis of stock levels, incorporating visual data from surveillance cameras and textual inventory records to optimize supply chain efficiency.

Healthcare

Healthcare stands to benefit immensely from multimodal AI, with its potential to transform diagnostic and treatment procedures. By integrating and analyzing medical records, patient verbal descriptions, and medical imaging, multimodal AI can assist in diagnosing diseases with increased accuracy and speed. Today, advanced, multimodal AI-powered surgical robots like STAR (Smart Tissue Autonomous Robot), developed by doctors at Johns Hopkins University, are able to perform complex intestinal surgical procedures with minimal doctor intervention.

Manufacturing

Multimodal AI technology has led to the development of a new breed of autonomous robots. For example, a general-purpose humanoid robot called Figure 1 can safely interact with its environment and work alongside people. Its AI technology allows it to learn new tasks and perform useful jobs using human-like dexterity. It can successfully perform monotonous manufacturing tasks, often resulting in human error due to fatigue or limited attention span.

Security

Moreover, multimodal AI has significant implications for security and surveillance, enhancing capabilities through facial recognition technologies (visual), voice recognition (audio), and threat analysis from textual communication. This integrated approach allows for a more comprehensive security strategy, identifying potential risks more accurately and efficiently.

However, the deployment of multimodal AI raises ethical considerations, including privacy concerns, the risk of biased algorithms across different data types, and the potential for job displacement. These challenges necessitate a thoughtful and balanced approach to the implementation of this technology in business practices.

In conclusion, businesses making innovative use of multimodal AI in their operations will maintain a decisive edge in today's rapidly advancing AI landscape to become tomorrow's industry leaders.

Chatbots in Business Processes

A chatbot is a computer program that can replicate and understand human speech and text in order to interact with humans through various digital devices. Use cases range from a website help chat to a personal assistant that can handle emails, phone calls, or business correspondence. ChatGPT by Open AI, Claude Anthropic, Google Gemini (formerly Bard), and Grok by xAI (formerly Twitter) are some of the most popular chatbots used today.

LLMs and multimodal AI technology have given rise to

advanced chatbots with the ability to perform a variety of tasks that were previously unavailable. Today, multimodal chatbots with vision and audio capabilities can respond to visual and audio prompts. This expanded capability can be leveraged in countless applications across various business applications.

Figure 1, the previously mentioned robot, combines multimodal AI technology and robotic technology, enabling it to process both visual and audio prompts and then respond autonomously. While in its infancy, this technology will certainly advance to give birth to a whole market of purpose-built robots designed to carry out a variety of specialized tasks. It won't be long before this technology will be available in kits for both consumer and commercial applications.

A popular application for chatbots, or bots as they are commonly called, is processing job applications. The goal of these complex programs is to sift through hundreds if not thousands, of applications and cover letters in order to identify the applicants who best match the job requirements. Because these bots automate the first screening phase, human HR professionals may dedicate more time and focus to evaluating the most qualified applications. This strategy improves recruiting efficiency by shortening the hiring process while simultaneously making candidate selection more precise and fairer.

Bots can also be used to summarize large documents such as legal contracts, scientific papers, or research studies. This use case will improve as the token size of bots increases. A token is the amount of text data that a bot can process and return in a session. For example, Claude Pro has a token limit of 200k, translating to about 500 pages of text. That is almost enough to get you halfway through Leo Tolstoy's War and Peace which is 1,225 pages.

The field of chatbots is vast and ever-evolving, which can make embracing this technology challenging. However, this is the foundational technology that AI is built on, which makes understanding it vital for business growth and sustainability. The key to learning this technology is mastering prompting. A prompt is a text query or a vocal phrase used to engage a chabot in performing a specific task. This is especially true in generative AI, where text prompts can generate images, videos, and music.

Today, you can create your own chabot or Agent to perform tasks tailored to meet your specific needs. For example, I created a bot with ChatGPT 4 called "Pathology Detective," designed to help me understand lab test results from my PCP and other medical specialists, as well as the side effects of the medication I was taking. I tried to publish the bot on the Open AI GPT Store but was denied as it didn't meet their guidelines. The following is the instructional text I used to build the bot.

"You have expert knowledge in human pathology based on online access to published clinical studies, medical journals, and scholarly content. Your primary role involves curating and summarizing all available medical and health-related information related to the user's inquiry. You have access to a vast online database of published health-related data and clinical studies, which you can reference to provide feedback using Wolfram Alpha or Wolfram Language. When discussing remedies, you always emphasize the importance of consulting healthcare professionals for personalized medical advice. Provide relevant images from Wolfram Alpha when available. Always include references to published studies, scholarly works, and articles in and outside the United States."

Here is an example of a prompt I submitted to
Pathology Detective:
How long does it take for the cancer drug Relugolix to leave your system?

Here is the response I received:
Relugolix is an oral gonadotropin-releasing hormone (GnRH) receptor antagonist used in the treatment of various conditions like prostate cancer, uterine fibroids, and endometriosis. The drug works by reducing the amount of certain hormones in the body.

For the elimination of relugolix from your system, it's important to consider its pharmacokinetic properties: Half-life: Relugolix has a half-life of approximately 25 hours.

This means that it takes about 25 hours for the concentration of the drug in the blood to be reduced by half.

Time to elimination: Generally, it takes about 5 to 6 half-lives for a drug to be completely eliminated from the body. Based on the half-life of relugolix, it would be expected to be largely eliminated from the body within 125 to 150 hours, or about 5 to 6 days.

These values can vary depending on individual factors like metabolism, age, kidney and liver function, and other health conditions. If you have specific concerns about how relugolix may affect you or its elimination process, it's best to discuss these with your healthcare provider.

Here is an example of a prompt used to create an editorial image for an article titled: Absolute Intelligence-- The Real Challenge:

Create an editorial image for the article title: Absolute Intelligence: The Real Challenge. The article shows how the brain's complex design is evidence of superintelligence and not evolution. Perhaps use a closeup image of a super computer chip with the words "Holy Bible" printed on it.

The potential uses of AI Agents in business are beyond the scope of this book and only limited by your imagination. What is critical to understand is that Agents are quickly infiltrating businesses and will soon be as common as printers and scanners.

The conclusion here is not to waste time exploring how your firm may benefit from AI Chatbots, Agents, and generative AI technology. Early acceptance and use of this technology will help to future-proof your business and assure long-term success.

Experiment with chatbots and prompting techniques to become acquainted with the technology and discover how it may be tailored to your company's requirements.

Chapter 3

Topics

Chapter 3

Implementing AI in Business: Strategies and Considerations

Securing employee trust and confidence is a vital step in successfully implementing an AI automation approach into a business workflow. Business executives must demonstrate convincingly how important this technology is to the company's future success and growth.

It is critical to help employees see and realize that AI automation is a tool for augmenting their work and ensuring the company's competitive advantage.

The formulation of an AI strategy begins with a clear definition of the organization's strategic business objectives. Executives must outline specific pain points or potential for leveraging AI, such as increasing customer service, automating supply chain operations, or innovating product development.

Identifying a certain area of focus is critical.
Once the goal is defined, the next step is assessing the organization's readiness to handle it. Given AI's dependency on data, having high-quality, accessible, and structured data is essential for successful adoption. Executives must assess their current data architecture, identify limitations, and devise a strategy for gathering, cleansing, and managing critical data.

Transitioning to AI automation requires the creation of a culture of cross-functional collaboration. Executives need to promote open dialogue and knowledge exchange, bridging business and technical teams. This inclusive approach ensures that the AI strategy is well-rounded, meeting business objectives while harnessing AI's technical advantages.

Additionally, understanding the ethical implications surrounding the use of AI is essential. With AI's growing influence, issues concerning privacy, bias, and accountability cannot be overlooked. Executives are tasked with setting up and developing comprehensive guidelines and frameworks to ensure AI's responsible use, emphasizing user privacy and fairness.

Continuous monitoring and adaptation are critical to the longevity of the AI strategy. AI technology and its applications are continually advancing; staying current with these advancements and understanding their ramifications is critical. Regularly examining performance indicators and obtaining stakeholder feedback will enable leaders to make educated decisions, promoting continuous improvement.

Finally, forging a successful AI strategy demands a comprehensive grasp of AI technology, clear alignment with business goals, readiness in data handling, collaborative effort, ethical governance, and relentless evaluation. By effectively communicating AI's role in propelling the company forward and gaining employee buy-in, business executives can leverage AI to spur innovation, operational efficiency, and sustainable growth within their organizations.

Analyzing Industry Challenges

Every industry has unique challenges and opportunities that can be effectively addressed through AI. The primary aim here is to uncover unique and novel applications of AI technology in your industry area. Remember that your industry experience and product knowledge may not be well-known among AI professionals.

As a result, there are several industries and vertical markets that have not fully embraced AI automation. If this is true in your case, early adoption of AI technology has the potential to catapult your business to a leadership position in your industry.

Do Market Research

To obtain a competitive advantage, it is important to keep up with the newest AI trends and advances in your sector. Explore case studies, white papers, and research articles to learn how your peers and competitors use AI. Attend AI-focused industry conferences, seminars, or webinars to network with professionals and learn from successful implementations.

Collaboration with AI Experts

While you may have a thorough understanding of your market, collaboration with AI experts can provide crucial assistance in developing AI use cases that are relevant to your company. Contact AI consultants, data scientists, or technology vendors who specialize in

your sector.
Use their experience to assess your current data infrastructure, data quality, and data availability to establish the viability and possible impact of AI efforts.

Testing with Beta Projects

Once you've identified potential AI use cases, start small with beta projects or proof-of-concepts to assess the feasibility and usability of your ideas. Choose a specific problem or process to optimize. Collaborate with internal teams or external partners to gather data, build AI models, and assess the outcomes. These small-scale trials will allow you to learn, iterate, and fine-tune your AI projects before implementing them throughout your organization.

Following these steps will help you identify AI use cases and potential in your industry. Remember that AI is not a one-size-fits-all solution; its potential impact will be defined by the characteristics and challenges specific to your industry. Accept AI's promise to drive innovation, stay ahead of the competition, and generate new revenue opportunities for your business.

🤖 🤖 🤖 🤖

Chapter 4

Topics

Data Collection and Preparation

Chapter 4

Data Collection and Preparation for AI Implementation

○────────────────────○

Data is the lifeblood of AI systems, laying the groundwork for training and decision-making. Without high-quality data, AI algorithms may deliver inaccurate or biased findings, reducing the effectiveness of your AI endeavors. As a result, it is critical to build strong data-collecting techniques and ensure the data is ready before going on an AI journey.

Data collecting entails acquiring useful information from a variety of sources, including customer interactions, operational procedures, and external databases. As a business leader, you must determine the essential data points that are relevant to your AI objectives and develop data-gathering strategies accordingly. This could include installing sensors, utilizing existing databases, or using data collection technologies. It is critical to establish a balance between the amount and quality of data, ensuring that the information gathered is comprehensive, reliable, and indicative of the mission at hand.

Following data collection, the next phase is data preparation, which entails cleaning, categorizing, and translating the data into a format suitable for AI algorithms. This approach demands careful attention because it has a substantial impact on the performance and dependability of your AI systems.

Data cleaning is removing any inconsistencies, duplicate entries, or missing values that may bias the results. Outliers and noise must also be discovered and treated so that they do not influence the AI model's behavior. Data organization entails categorizing the data into relevant categories or properties that AI algorithms can recognize. This could include data labeling or annotation, allowing the AI system to identify trends and make accurate predictions.

Data transformation is turning data into a format that AI algorithms can handle efficiently. Depending on the technique, this could entail scaling, normalizing, or encoding the data. It is critical to ensure data integrity and that the transformation process does not introduce bias or distort the original information.

Business executives may create a solid foundation for effective AI deployment by putting in the time and resources to collect and prepare proper data. High-quality data enables AI systems to provide accurate insights, automate processes, and make intelligent judgments. It is a critical step in helping corporate executives understand AI and properly exploit this game-changing technology.

Finally, company executives must collect and prepare data as part of the AI deployment process. Executives may realize the full potential of AI technology by recognizing the importance of data quality, developing robust data-gathering strategies, and assuring data preparation through cleaning, organizing, and converting processes.

Adopting AI with a solid data foundation enables businesses to acquire a competitive advantage, accelerate innovation, and achieve long-term success in a quickly changing digital market.

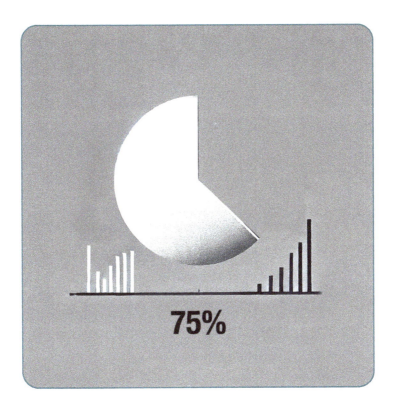

Topics

Data Quality and Availability

Chapter 5

Data Quality and Access

———————————○———————————

The quality and availability of data are critical to the success of artificial intelligence programs. AI initiatives can fail due to a lack of accurate, reliable, and accessible data, resulting in poor results and missed opportunities. In this chapter, we will discuss the significance of assuring data quality and availability for AI projects, giving business executives a thorough understanding of the essential aspects.

AI projects require high-quality data. Inputting low-quality data into an AI system can result in inaccurate results and skewed conclusions. As a business executive, you must implement strong data governance policies to assure data correctness, completeness, and consistency. This includes developing data validation methods, conducting regular data audits, and utilizing data sanitizing techniques to reduce errors and inconsistencies that could ruin AI initiatives.

Furthermore, data availability is critical to the success of artificial intelligence programs. Without access to relevant and timely data, AI systems are unable to create meaningful insights or make correct predictions. As a business executive, you must have a well-defined data strategy that covers data collection, storage, and retrieval. This includes finding the appropriate data sources, deploying data integration solutions, and creating data pipelines that allow for continuous data flow to AI models.

Collaboration between business and IT teams is critical for maintaining data quality and availability. Business executives should collaborate closely with data scientists, analysts, and IT professionals to define data requirements, identify potential data gaps, and set up data governance rules. Organizations may ensure that AI projects are built on solid data by cultivating a collaborative and communicative culture.

Additionally, improvements in AI technology, such as automated data quality checks and machine learning algorithms for data cleaning, can help company executives maintain data accuracy and availability. Using these tools, organizations may automate the discovery and resolution of data issues, freeing up important time and resources for other critical tasks.

Osmos.io is one SaaS firm that offers data cleaning and migration services to help streamline data integration into AI models.

In conclusion, assuring data quality and availability is critical to the success of AI projects in the commercial world. To ensure data integrity, dependability, and accessibility, business executives must prioritize data governance, develop strong data policies, encourage collaboration between business and IT teams, and employ advanced data gathering and cleansing technology. This allows enterprises to fully realize the potential of AI and achieve a competitive advantage in today's data-driven economy.

Chapter 6

Topics

Ethical and Privacy Concerns in AI adoption

Chapter 6

Ethical and Privacy Concerns in AI Adoption

Artificial intelligence (AI) has emerged as a game changer for corporations across sectors. It has transformed the way businesses work, streamlining procedures, increasing consumer experiences, and accelerating innovation. However, the rapid progress of AI technology has brought significant ethical and privacy problems that must be addressed. Business leaders must address these concerns to enable responsible and sustainable AI deployment.

One of the key ethical concerns with AI is the possibility of bias in decision-making algorithms. AI systems learn from massive volumes of data, and if that data is skewed or incomplete, they can produce discriminatory results. This can have major ramifications for firms, including reputational harm and legal action. To address this issue, business executives should prioritize fairness in AI algorithms by conducting regular audits, using diverse and inclusive datasets, and integrating ethics experts in the development process.

Another area of concern is the impact of artificial intelligence on jobs and the workforce. While AI can automate tedious work and increase productivity, it also raises concerns about job displacement. Business executives must address these issues proactively by investing in staff reskilling and upskilling initiatives, guaranteeing a smooth transition to an AI-powered workforce. They should also communicate openly with employees, reassuring them about the importance of human skills and the new opportunities that AI can provide.

Privacy is another major issue in the use of AI. AI systems rely on massive volumes of personal data, so there is a risk of data breaches and unauthorized access. Business executives must prioritize data protection by establishing strong cybersecurity measures according to relevant rules such as the General Data Protection Regulation (GDPR) and being open and honest with customers about how their data is acquired, utilized, and safeguarded.

Additionally, ethical issues should apply to the employment of AI in high-risk industries like healthcare, banking, and criminal justice. Business leaders must guarantee that AI systems are used properly in accordance with legal and ethical standards. They should create clear norms and governance structures, evaluate the impact of AI systems on a regular basis, and include stakeholders in decision-making processes.

To summarize, while AI provides huge benefits for organizations, addressing ethical and privacy concerns is critical. Business leaders must be proactive in assuring justice, openness, and accountability in AI adoption. This allows them to create trust with customers, employees, and society while also realizing the full potential of AI technology for their organizations.

Chapter 7

Topics

Cultivating AI Talent and Capabilities within Your Organization

Chapter 7

Cultivating AI Talent and Capabilities within Your Organization

Developing AI expertise is critical for firms beginning their AI journey. This includes finding individuals within the organization who have the talent or capacity to study AI. Executives may develop AI talent internally by instilling a culture of continuous learning and giving opportunities for upskilling. Additionally, corporations can attract external talent by providing competitive remuneration packages and collaborating with academic institutions to gain access to AI knowledge.

Creating a specialized AI team or center of excellence (CoE) is another efficient way to develop AI skills. This team consists of data scientists, AI engineers, and domain specialists who work together to create and deliver AI solutions. The AI CoE serves as a focus for knowledge sharing, experimentation, and innovation, ensuring that AI projects are in line with corporate goals.

Cultivating AI talent in-house is an excellent strategy for establishing an AI CoE because it lowers personnel acquisition expenses and eliminates the time required to learn your industry.

To encourage AI adoption, leaders must establish a data-driven culture in their firms. This includes investing in a strong data infrastructure, assuring data integrity and security, and encouraging data literacy among staff. Organizations may empower individuals to make informed decisions and drive AI-driven initiatives by democratizing data access and providing employees with the tools and skills they need to evaluate and interpret data.

Finally, CEOs should actively seek out relationships with external AI vendors and startups. These collaborations can give enterprises access to cutting-edge AI technologies and experience, helping them to speed up their AI initiatives and solve implementation obstacles.

To summarize, cultivating AI talent and capabilities within an organization is critical for business executives seeking to realize the potential of AI technology. Executives can put their firms at the vanguard of the AI revolution by investing in people development, establishing specialized AI teams or Centers of Excellence, cultivating a data-driven culture, and cultivating strategic relationships. Embracing AI not only allows firms to improve operational efficiency and customer experiences but also creates new opportunities for development and innovation in an increasingly competitive business environment.

Chapter 8

Topics

Managing Change - Considerations

Managing Change - Implementation

Chapter 8

Managing Change and Resistance to AI Adoption

———————————————————

Change is a natural element of life. Implementing AI within an organization requires a shift in mindset, processes, and culture. Effective change management is critical to ensuring a smooth transition and maximizing the benefits of AI adoption. This chapter delves into various strategies and best practices for managing change, including:

Considerations

1. Creating a convincing vision for AI adoption that connects with overall business goals is critical. Communicating this vision to stakeholders will help generate enthusiasm and support for AI initiatives.

2. Developing an AI-ready culture can help ensure successful AI implementation. The formation of an AI center of excellence (CoE) is an excellent way to demonstrate to stakeholders the organization's firm commitment to AI adoption and the benefits it offers to company growth. Encouraging employees to see AI as a tool to improve their work rather than a threat to their careers is critical in overcoming reluctance.

3. It is critical to provide thorough training programs to staff so that they are equipped to work with AI technologies. This will assist in relieving worries and increasing confidence in using AI tools efficiently. Resistance to change is a prevalent issue during any organizational shift. Many company executives may be hesitant to fully adopt AI for a variety of reasons, including job security, privacy, and ethical considerations. This sub-chapter discusses these concerns and offers techniques for overcoming opposition.

Implementation

1. Realistic Expectations: Corporate leaders must establish realistic expectations for staff output efficiency and productivity levels while adopting AI. There will be a steep learning curve at first, but efficiency and production will improve over time as employees become more AI-competent.

2. Transparent Communication: Addressing issues openly

and offering clear, factual information regarding AI can help employees overcome their worries and create trust. Communicating the advantages of AI adoption, such as enhanced efficiency and better decision-making, may help ease opposition.

3. Collaboration and Involvement: Involving employees in decision-making and soliciting their feedback helps promote a sense of ownership while decreasing opposition. Creating cross-functional teams to oversee the AI deployment process can also help to guarantee that different perspectives are taken into account.

4. Begin Small, Scale Fast: Using AI in small, manageable initiatives enables testing and learning without disturbing the entire organization. Early successes can boost confidence and inspire support for wider AI usage.
By embracing AI technology and effectively managing change, business executives can unlock its vast potential to drive innovation, improve efficiency, and gain a competitive edge. This sub-chapter provides practical insights and strategies to help business executives navigate the complexities of AI adoption, making it simple and accessible for executives from all backgrounds.

Chapter 9

Topics

Evaluating ROI and Measuring Success

Chapter 9

Evaluating ROI and Measuring Success

When calculating ROI, it is critical to evaluate both the concrete and intangible benefits that AI may provide to your firm. Cost reductions, higher revenue, improved efficiency, and less human error are all examples of tangible benefits. Intangible advantages, on the other hand, include things like improved customer experience, better decision-making, and greater agility in responding to market changes.

To assess the success of AI projects, create explicit objectives and key performance indicators (KPIs) from the start. These KPIs should be consistent with your business objectives and give measurable results. For example, if your goal is to improve customer satisfaction, you may assess AI's effectiveness in lowering customer complaints or increasing positive feedback.

To accurately evaluate ROI, relevant data must be collected and analyzed. Data might come from different sources, such as consumer interactions, sales numbers, operational analytics, or employee feedback. Collecting and analyzing data can help you understand the impact of AI on your organization and make informed decisions.

Another crucial part of determining AI success is to assess the long-term rewards and hazards. While AI might deliver immediate benefits, it is equally critical to consider its possible impact on your business in the future. This involves recognizing potential scalability concerns, the necessity for continuing maintenance and updates, and the possibility of ethical or legal consequences.

Furthermore, communication between corporate executives and AI professionals is critical for determining AI success. Engaging with AI professionals and asking their advice will help you understand the intricacies of AI adoption and ensuring that your review is thorough.

To summarize, calculating ROI and measuring AI success is critical for corporate executives seeking to maximize the potential of AI technology. To make informed decisions and optimize the benefits of AI, it is vital to consider both real and intangible benefits, setting clear objectives and KPIs, analyzing relevant data, and assessing long-term implications. Engaging with AI professionals and exploiting their knowledge is also critical to success in this field. Taking these steps will help ensure that AI will help your business achieve both short-term and long-term objectives.

♔ ♔ ♔ ♔

Chapter 10

Topics

Real-world Examples of AI Success in Business

Real Estate

Banking

Healthcare

Education

Chapter 10

Real-world Examples of AI Success in Business

AI-powered solutions are transforming how organizations address a variety of business objectives by using algorithms and machine learning techniques to filter through massive volumes of data. This technology automates and streamlines monotonous work but also allows organizations to provide more tailored experiences, which is critical in today's competitive landscape. By providing real-time information, AI enables CEO's to make more educated decisions, improve operational efficiency, increase consumer engagement, and grow revenue.

The following are real-world examples of organizations that have successfully employed AI technology to help them problem-solve, expand, automate decision-making activities, and promote customer interaction.

REAL ESTATE

Redfin, a real estate agency, employs AI and machine learning to improve its services and add value to both buyers and sellers. Here's how Redfin successfully integrated AI.

1. Automated Home Valuations (Redfin Estimate)

Redfin's "Redfin Estimate" is a tool that estimates the market worth of homes. To provide accurate property appraisals, this AI-powered tool employs machine learning algorithms that assess a variety of data points, such as similar home sales, public records, and market trends. Redfin says that their assessments are more accurate than those of typical automated valuation methods because they learn and adapt when fresh data is available.

Result:
a) Provides home-buyers and sellers with a reliable assessment of a home's value, allowing them to make informed purchasing and selling decisions.

b) Help democratize sales in the real estate industry by providing customers with a tool to compare house prices and analyze market patterns.

2. Personalized Home Search

Redfin employs artificial intelligence to tailor the home search experience for its users. The software analyzes user behavior, such as search history, saved houses, and viewed listings, to suggest properties that are likely to meet the user's needs. To enhance its recommendations, the AI evaluates parameters such as location, price range, and home attributes.

Result:
a) Helps consumers identify residences that better meet their needs, minimizing the time and effort
spent searching.

b) Increases user engagement by providing a more tailored and efficient search experience.

3. Market Insights and Analytics

Redfin uses AI to generate real estate market insights and analytics. The platform analyzes massive datasets, including home prices, sales patterns, and demographic data, using machine learning to provide customers with detailed insights and forecasts. This data assists home-buyers and sellers in better understanding market conditions and making more educated decisions.

Result:
Users gain access to advanced market analysis that would normally need a real estate professional, allowing them to make smarter decisions.

Improves Redfin's reputation as a tech-forward real estate platform with unique data insights.

4. Streamlined Operations

Redfin has incorporated AI into its back-office operations to increase efficiency. AI is utilized to improve agent efficiency, maintain customer engagement, and automate regular tasks like scheduling and follow-ups. This enables Redfin agents to concentrate more on providing personalized service to clients.

Result:
a) Increased operational efficiency, lowered expenses, and improved service delivery.

b) Allows Redfin agents to handle more clients and transactions, resulting in increased business development.

5. Enhanced Customer Service

Redfin uses AI-powered chatbots and virtual assistants to handle customer inquiries and provide instant support. These AI tools can answer common questions, guide users through the home buying or selling process, and even help schedule appointments with agents.

Result:
Gives users instant support, enhancing the overall customer experience. Allows human agents to focus on complex tasks, resulting in more efficient service. Redfin's adoption of AI into many facets of its business has enabled it to provide more accurate, tailored, and efficient services, propelling it to the forefront of the real estate industry's technological transition.

BANKING

JP Morgan Chase has developed various AI-driven projects to improve operations, risk management, and customer service. Here's a closer look at how they've used AI:

1. Contract Intelligence (COiN)
One of JP Morgan's most significant AI deployments is

the COiN (Contract Intelligence) platform. COiN analyzes and extracts information from legal documents such as loan agreements and contracts by utilizing machine learning
and natural language processing (NLP). This system greatly shortens the time needed to review and process these documents.

Result:
a) Previously, it took 360,000 hours annually to review legal documents. COiN performs the same task in just seconds.

b) COiN improves accuracy by reducing human errors that are typical in manual document assessments, particularly when dealing with huge amounts of data.

c) It enables legal and compliance personnel to focus on more complicated, value-added work rather than basic document analysis.

2. Fraud Detection and Risk Management
JP Morgan uses AI to enhance fraud detection. The bank uses machine learning algorithms to examine transaction trends and identify anomalies that may suggest fraudulent activity. The AI system is constantly learning from new data, improving its capacity to detect even minor signals of fraud.

Result:
a) Enhanced fraud detection accuracy, leading to a significant reduction in financial losses due to fraudulent activities.
b) The ability to handle and analyze large amounts of transaction data in real-time, resulting in faster reactions to potential fraud.

3. Personalized Banking Services

JP Morgan has also investigated AI in the context of personalized banking. Artificial intelligence-powered systems analyze consumer data to provide individualized financial advice, product recommendations, and tailored services. For example, the bank's artificial intelligence-powered virtual assistant assists customers in managing their accounts, answering questions, and even providing investing ideas based on their financial patterns.

Result:
a) Improved client satisfaction by providing more personalized and relevant service options.

b) Increased efficiency in answering common client inquiries, allowing human agents to focus on more complicated issues.

4. AI in Trading

JP Morgan is using AI in its trading operations to improve trade execution and identify market trends. The AI algorithms examine market data, news feeds, and other pertinent information to forecast market moves and recommend the best trading methods. This has propelled the bank to the forefront of algorithmic trading.

Result:
a) Increased trading efficiency and profitability by executing transactions using real-time data and powerful AI-powered analysis.

b) Improved decision-making in volatile markets leads to lower risk.

5. Operational Efficiency
JP Morgan also uses AI to automate back-office activities like data input, compliance checks, and customer service routines. Robotic Process Automation (RPA) is used to manage repetitive activities, providing consistency while eliminating the possibility of human error.

Result:
a) Automating routine, repetitive operations resulted in significant cost reductions.

b) Increased operational efficiency, resulting in faster service delivery and more accurate administrative processes.

JP Morgan's usage of AI extends to numerous aspects of business operations, highlighting AI's disruptive impact on traditional banking processes. The bank continues to innovate, embracing artificial intelligence to remain competitive in today's highly volatile financial world.

HEALTHCARE

DEEP6 AI is a healthcare technology business that employs artificial intelligence to transform patient recruiting for clinical trials. Clinical trials are critical for moving medical research forward, but locating qualified participants can be time-consuming and difficult. Deep 6 AI tackles this problem by using AI to quickly identify eligible patients from massive amounts of medical data.

1. Data Integration and Analysis
Deep 6 AI processes and analyses a diverse set of structured and unstructured data sources, including electronic health records (EHRs), clinical notes, diagnostic reports, and lab results. Natural language processing (NLP) is used in the platform to extract essential information from unstructured data, such as doctor's notes, which frequently contain important facts about a patient's medical history.

Result:
a) Faster Recruitment Speed: Deep 6 AI has greatly

decreased the time required to locate qualified applicants for clinical trials. What once took months or even years can now be completed in weeks or days.

2. Patient Matching

DEEP6 AI sorts through the data to find patients who meet specified clinical trial criteria. For example, if a trial requires volunteers with a specific medical condition,
the AI can swiftly identify individuals who meet these requirements, even if the information is buried in unstructured text.

Result:
a) Enhanced Matching Accuracy: Deep 6 AI can identify eligible patients with better accuracy than previous approaches, ensuring that more suitable applicants are chosen for trials.

b) Improved Trial Efficiency: Faster and more accurate recruiting methods result in faster trial completions, which accelerates medical research and the development of novel medicines.

c) Expanded Participant Pool: The ability to sift through big datasets enables the identification of eligible participants who would have been ignored using traditional approaches, hence increasing the pool of possible trial applicants.

3. Continuous Learning

As the AI processes more data and learns from user feedback, it continuously improves its accuracy in identifying suitable candidates.

Result:
a) Greater Accuracy: The self-learning characteristics of the system make it more effective over time, making the recruitment process faster and more dependable.

Deep 6 AI illustrates the revolutionary potential of AI in the healthcare industry. By automating and improving the patient recruitment process, it not only speeds up clinical trials but also increases the overall efficiency of medical research, resulting in faster creation of new treatments and therapies.

EDUCATION

Georgia State University (GSU) struggled with high dropout rates and low graduation rates. To solve this issue, GSU integrated an AI-powered predictive analytics technology into their student advising process. The AI system, known as the "Panther Retention Grant Program," analyzes student data to identify those who are at risk of dropping out and intervenes to provide assistance.

1. Predictive Analytics

This AI system employs predictive analytics to track over 800 characteristics per student, including class attendance, grades, and financial aid status. The algorithm recognizes trends that may signal a student's risk of academic failure or dropout.

Result:
a) Increased Graduation Rates: Since the deployment of AI, GSU's graduation rates have grown dramatically. Over the course of a decade, the university's graduation rate rose from 32% to 54%, thanks in large part to the AI-driven initiative.

2. Automated Alerts

When the AI senses that a student is struggling, it instantly notifies academic advisors, who may then help early. Counseling, tutoring, and course load changes are examples of such interventions.

Result:
a) Improved Retention: The usage of AI has resulted in

higher retention rates, particularly among students from traditionally undeserved backgrounds. Interventions tailored to individual student needs help them stay on track by addressing difficulties before they become dropouts.

b) Cost Effectiveness: This cost-effective, proactive approach has reduced the need for more expensive remediation later in a student's educational career. This efficiency has enabled GSU to support a bigger number of students with fewer resources.

GSU's use of AI in academic counseling exemplifies how predictive analytics can create significant gains in educational results. By targeting at-risk students early and delivering tailored interventions, the university was able to drastically boost both retention and graduation rates, highlighting AI's problem-solving potential in education.

Conclusion:
These are just a few instances of businesses and organizations that have embraced AI technology and are now experiencing its benefits. Further breakthroughs in AI will broaden its application in a wide range of scenarios. Many of these developments will come from within industries and enterprises that have discovered unique ways to employ AI to meet their specific demands. The message here is to examine how AI is being used in

your market area and determine how your firm can adopt and innovate on it.

        👑   👑   👑   👑

Chapter 11

Topics

AI Agents

Smarter Smart Phones

More Powerful Processing

The Brains behind AI

Chapter 11

The Future of AI in Business

○──○

      AI is evolving at an alarming rate, making it difficult, if not impossible, to predict its future in business. As open-source AI advances, enterprising developers will flood the market with new AI agents, smart devices, personal robots, and more. This ebb and flow will continue until the market achieves parity. When a market reaches parity, in which competing products or services are almost equivalent in terms of quality, features, and price, businesses may increase competition through aggressive marketing, branding, and customer experience activities. Innovation is required to stand out, but the market may become commoditized, resulting in price sensitivity and potential price wars. To avoid direct competition, businesses may seek niche sectors, while customer loyalty becomes increasingly crucial in retaining market share.

One thing is certain, businesses that fail to adopt AI technology will be unable to compete with rivals who have implemented the technology. The following is a list of innovations in AI technology and leading companies that are shaping the future of Artificial Intelligence. Some of these innovations will hit the market before this book is published and many more new AI products and services will emerge as well.

AI Agents

When Chatbots first appeared, consumers needed an internet connection and a web browser to communicate with them. Currently, larger model LLMs are used to train smaller LLMs. This allows smaller variants to run on local machines as well as smart devices. For example, OpenAI and Apple have collaborated to integrate ChatGPT into Apple's operating systems, including iOS, iPadOS, and macOS.

This agreement, announced at Apple's 2024 Worldwide Developer Conference, gives customers direct access to AI-powered capabilities, including better image and document comprehension within Apple's ecosystem. Siri can improve user interactions with devices by leveraging ChatGPT (OpenAI).

The integration will also include Apple's Writing Tools, which will allow users to create text and employ image-generating tools in a number of styles that are immediately linked into their writing processes.

Importantly, privacy safeguards are taken into account; interactions with ChatGPT via Siri and Writing Tools do not log user requests, and IP addresses are masked to protect privacy. Users can personalize their experience by connecting their ChatGPT accounts, but this is not required to use basic functionality.

Smarter Smart Phones
Samsung's Android interface, One UI, features AI technology in the Galaxy S22 series and above, improving the functionality and user experience of Samsung phones. This integration offers advanced multitasking features like resizable picture-in-picture and a smoother Samsung DeX experience, bringing the mobile environment closer to desktop capabilities.

The interface also allows for significant customization with features like Smart Widget, which dynamically alters the information presented based on user activities, as well as security measures like passkeys and an auto-blocker to improve device security.

The system also focuses on user health and accessibility, with tools for managing screen time and thorough feedback on digital behaviors to encourage more conscious use. Accessibility features have been enhanced, providing better visual aids and magnification options for people with impaired vision. Samsung Health is smoothly integrated with the UX, adding health tracking and goal-setting features to help users maintain their heath. With these features, One UI seeks to provide a more personalized and secure user experience throughout Samsung's contemporary smartphone lineup.

More Powerful Processing
Intel, one of the leading computer chip manufacturers, is actively incorporating AI processing into its product designs, particularly with its latest Meteor Lake processors. These chips are intended to improve performance and efficiency by lowering power consumption, increasing processing speeds, and more effectively managing workloads. AI capabilities are integrated across the chip-making process, from design and production to software development, offering increased yields and improved performance.

Intel's approach includes the development of processors capable of handling AI workloads in a variety of contexts, including data centers and edge devices. This will enable their AI-capable chips to handle a wide range of AI processing needs. For example, Intel processors handle large-scale AI workloads in data centers, whereas edge computing brings AI capabilities closer to where data is created, which is critical for applications such as autonomous vehicles and smart cities.

Intel is currently researching breakthrough technologies such as neuromorphic and quantum computing to further improve AI processing. Neuromorphic computing replicates the neural networks of the human brain, which has the potential to speed up and save energy in AI computation. Quantum computing, while still in its early stages, provides the possibility of solving challenging AI issues beyond the capabilities of traditional computers. These improvements will have a substantial impact on a variety of industries, including consumer technology, healthcare, automobiles, and finance.

Qualcomm, a key provider of chips for smartphones and other smart devices, is incorporating AI processing into its chip designs. This will significantly expand the capabilities and efficiency of these devices. Qualcomm's Snapdragon X Elite platform greatly improves AI processing to enhance both performance and power efficiency, which is critical for devices that require high computational power while maintaining battery life. This platform enables more complex AI operations to be performed directly on devices, expanding apps such as augmented reality and real-time language translation, resulting in a more fluid and responsive user experience.

These hardware breakthroughs give developers access

to a library of pre-optimized AI models through Qualcomm AI Hub. This program aims to standardize the deployment of AI capabilities across devices, making it easier for manufacturers to integrate advanced AI features into their products. For customers, this means smartphones and other devices will be able to provide more personalized services by better adjusting to user preferences and behaviors. Camera applications, for example, may automatically modify settings based on the shooting environment, while digital assistants could deliver more precise responses based on context.

Qualcomm's AI-enhanced chips have far-reaching consequences beyond smartphones, affecting a diverse spectrum of smart devices in the IoT and automotive sectors. These chips in IoT could enable smarter home automation systems that better anticipate consumer wants. In the car business, advanced AI processing could improve navigation systems, safety features, and vehicle communications. Qualcomm's solution solves security and privacy concerns by processing data locally on devices, removing the need to send sensitive user data to the cloud. This comprehensive enhancement in device intelligence, efficiency, and security represents a huge step forward in the general use of AI across multiple technological fields.

nvidia is a key player in the advancement of AI technology today, primarily through the development and sale of Graphics Processing Units (GPUs), which are required for AI calculation. NVIDIA GPUs are well-known for their ability to handle parallel operations efficiently, making them perfect for the computationally intensive processes required to train and execute AI models, particularly deep learning and neural networks.

Aside from hardware, NVIDIA provides a comprehensive software stack, which includes CUDA, a parallel computing platform and programming model that greatly improves computing speed by leveraging GPUs. This allows developers to construct and deploy a wide range of AI-powered applications more efficiently. This stack includes NVIDIA's cuDNN library, which provides finely tuned implementations of standard techniques such as forward and backward convolution, pooling, normalizing, and activation layers.

NVIDIA's AI platform is more than simply a toolkit for developing AI solutions; it also includes applications such as autonomous vehicles, robotics, and high-performance computing. For example, their DRIVE platform is utilized to build self-driving technologies, while their Jetson platform is popular for robotics and embedded applications. These integrations demonstrate how NVIDIA's technology is not only fundamental to AI research and development but also essential for applying AI to real-world challenges. NVIDIA's activities keep it at the forefront of the AI technology environment, fostering innovation in a variety of industries.

apple has established itself as a prominent player in the AI business with its recent collaboration with OpenAI. This collaboration enables users to use ChatGPT via Siri and other system-wide features on iOS, iPadOS, and macOS. The connection will be available with iOS 18, iPadOS 18, and macOS Sequoia.

Users will be able to use ChatGPT for a range of tasks, including text generation, image creation, story composition, and detailed solutions to challenging queries via Siri. Siri, for example, may recommend utilizing ChatGPT to aid with food planning, bedtime reading, or creative writing.

One of the primary benefits of this integration is increased privacy. Apple has underlined that the collaboration includes strong privacy safeguards, such as anonymizing user IP addresses and assuring that OpenAI does not store user requests. Furthermore, users will be able to use ChatGPT for free, with the option to link their ChatGPT Plus accounts for premium services.
This relationship benefits both companies.
Apple improves Siri's skills, making its gadgets more versatile, and OpenAI obtains visibility to millions of Apple customers, perhaps increasing subscriptions to its premium services.

Sidebar:

Apple's M2 and M3 CPUs have notable performance benefits compared to Windows platforms with separate GPUs. This is because of their Unified Memory Architecture, which decreases latency and improves efficiency by enabling the CPU and GPU to utilize the same memory. Apple's ecosystem achieves enhanced performance and battery efficiency through the seamless integration of hardware and software, particularly in demanding applications such as video editing and machine learning. Moreover, Apple's specialized accelerators and energy-efficient architecture offer exceptional performance per unit of power, making these chips especially suitable for mobile devices. This contrasts with the higher power usage and potential inefficiencies found in conventional Windows systems with separate graphics processing units (GPUs).

The Brains Behind AI

baidu, widely known as the "Google of China," developed Baidu Brain, an AI platform that combines natural language processing, image recognition, and deep learning and is at the heart of the company's AI activities. This platform serves as the foundation for the bulk of Baidu's products and services, enhancing both user experience and operational efficiency.

Baidu's Apollo is an open-source autonomous

vehicle platform that works with worldwide automotive manufacturers to improve self-driving technology. It is regarded as one of the world's leading programs in autonomous driving, making important contributions to the evolution of smart transportation systems. Additionally, Baidu invests substantially in AI-powered health technologies, such as AI algorithms, that aid in medical imaging and diagnostics, hence improving the accuracy and efficiency of healthcare services. These activities illustrate Baidu's commitment to leveraging AI for societal benefit, driving innovation in vital areas, including healthcare, mobility, and digital services.

claude AI was developed by Anthropic, a technology company that specializes in advanced AI systems. Claude is a computer program that uses natural language interaction to conduct conversations, answer inquiries, and aid with various activities.

Established in 2021, Anthropic's aim is to create safe and ethical AI systems that benefit humanity. They are well-known for their work on AI alignment, which seeks to construct AI systems with goals and actions that are consistent with human values.

Claude is designed to aid with a variety of tasks, such as analysis, writing, coding, problem-solving, and more. Claude has access to a large knowledge library spanning a wide range of topics, but it lacks real-time information and the capacity to learn and update its knowledge through discussions. Claude's responses are generated using patterns from the data on which it was trained (up to April 2024), with the goal of providing relevant and contextually suitable information.

Google **bard** is a conversational AI model built on Google's Language Model for Dialogue Applications (LaMDA). It's designed to interact in a conversational fashion, offering responses that are built on a large amount of information from the internet. Bard strives to aid users with information search, creative content development, and addressing queries with a more personal touch. It's coupled with real-time data from the web, allowing it to deliver current information, which differentiates it from certain other models like OpenAI's ChatGPT that are trained up to a fixed point in time.

Google **gemini** is a larger, more ambitious multimodal AI model, which means it can interpret and generate information across diverse sorts of data, including text, images, audio, and video. Gemini is part of a bigger ecosystem that is endeavoring to incorporate AI more deeply into various Google goods and services. It's developed by Google DeepMind and is designed to execute difficult reasoning tasks, which can be deployed across a range of Google's products and third-party tools. Gemini comes in multiple versions (Ultra, Pro, and Nano) adapted to diverse operating demands, from data centers to mobile devices.

Gemini and Bard are both part of Google's AI programs; Gemini is a more comprehensive, multimodal AI platform that supports a wide range of applications, while Bard focuses on improving users' conversational experiences by offering contextually aware and up-to-date information from the web.

ibm, best known for Watson, which debuted on the TV quiz show "Jeopardy" in 2011, has been a pioneer in AI since the early 1950s. Its most recent AI projects have focused on IBM Watsonx, an AI and data platform designed to help enterprises get insights, increase efficiency, and improve customer experiences. This platform integrates a number of AI tools and services, making it easier for organizations to design AI solutions that are tailored to their specific needs.

In 2023, IBM established a $500 million Enterprise AI Venture Fund to invest in AI startups, with a specific emphasis on generative AI technology and research. This fund is meant to encourage creativity and promote the development of new AI applications, keeping IBM at the forefront of AI breakthroughs. By investing in startups, IBM is forming strategic alliances that will shape the future of AI technology.

Another noteworthy endeavor is IBM's partnership with

NASA to create the largest geospatial foundation model on Hugging Face, an open-source AI collaboration platform. This approach is intended to assist academics and developers in creating AI systems that can analyze and interpret geospatial data, which has wide applications in domains such as environmental science, urban planning, and disaster response.

IBM has also been actively encouraging responsible AI development. IBM formed the AI Alliance with Meta and other prominent organizations to promote open, safe, and responsible AI. This program seeks to establish best practices and guidelines for AI development, ensuring that AI technologies are used ethically and responsibly. Meta (Meta Platforms, Inc.) is a worldwide technology conglomerate located in Menlo Park, California. Meta, formerly known as Facebook, Inc., was founded by Mark Zuckerberg, the current chairman and CEO. The corporation was relaunched in October 2021 to reflect its goal of creating the metaverse, an integrated ecosystem that connects its goods and services.

meta owns and administers several prominent platforms, including Facebook, Instagram, Messenger, Threads, and WhatsApp. While Meta is not an open-source project, the firm contributes heavily to open-source AI development, most notably with Llama 2, a sophisticated large language model distributed freely for study and commercial usage in 2023. React, a popular

JavaScript library for creating user interfaces, is another one of Meta's open-source initiatives.

Meta's offerings are grouped into two categories: family of apps and reality labs. Facebook, Instagram, Messenger, and WhatsApp are part of the Family of Apps section, which allows individuals to interact and share using numerous modes of communication. Reality Labs specializes in augmented and virtual reality products, including consumer gear, software, and content. OpenAI, created in December 2015, is an AI research company that began as a non-profit organization with the goal of freely sharing its findings, tools, and technologies to benefit all of humanity. The company has now transitioned to a for-profit corporation to better support its research objectives. OpenAI's purpose is to ensure that artificial general intelligence (AGI)—highly autonomous systems that outperform humans in the most economically useful tasks—benefits all of humankind. The company is well-known for its commitment to publicly sharing its research, conclusions, and technology with the rest of the world. However, it has modified its sharing policies to ensure safety.

openAI has created various notable AI models, such as the GPT (Generative Pre-trained Transformer) series. These models have made significant advances in natural language processing and understanding, displaying impressive skills to generate human-like prose based on prompts. GPT-3 and GPT-4, two of OpenAI's most prominent releases, are among the most powerful language models ever constructed, and they are widely used in a variety of applications ranging from automated content generation to complex decision support.

Aside from language models, OpenAI is well-known for its breakthroughs in robotics, AI safety research, and the creation of DALL-E, an AI system that can generate images from textual descriptions. This mix of projects exemplifies OpenAI's comprehensive approach to AI development, which focuses not only on practical applications but also on basic research that could lead to a better understanding and more resilient AI systems. OpenAI's alliances, such as its relationship with Microsoft and Apple, demonstrate its influence and the commercial potential of its innovations.

X.AI, led by Elon Musk, aims to use artificial intelligence to improve our understanding of the cosmos and speed up human scientific discoveries. This ambition is represented in the development of Grok, a conversational AI model that interacts with people and assists with scientific and general inquiries by offering detailed and knowledgeable responses. X.AI's innovative AI models, including Transformer-XL and Memorizing Transformer, improve their ability to handle complicated tasks and enormous datasets.

grok AI is a sophisticated conversational AI model developed by X.AI, a company led by Elon Musk. This AI is designed with the ambition of accelerating human scientific discovery and expanding our collective understanding of the universe. Grok AI utilizes a 33-billion parameter-dense transformer architecture, which allows it to handle complex tasks and manage large datasets effectively. The model was initially released in an early-access format as Grok-0 and has since been updated to Grok-1 and Grok-1.5, with each iteration introducing enhancements in its capabilities.

The Grok AI platform is integrated within X's services, offering users an interactive AI experience that can be utilized for both general inquiries and specialized scientific questions. The AI model has been involved in the development of several groundbreaking AI methods, such as Transformer-XL and the Memorizing Transformer, which have contributed to its advanced processing abilities. These features make Grok a powerful tool for a range of applications, from engaging with users in everyday conversations to tackling more complex, data-intensive tasks.

As previously stated, these are the key players who are now pushing the boundaries of AI development. The following is a short selection of companies that have established a niche in the AI market and should be on your radar.

Cubicasa

Cubicasa, headquartered in Oulu, Finland, is a mobile app that enables users to create detailed 2D and 3D floor plans by scanning a property with their smartphone. It is commonly used in real estate for marketing, appraisals, and listings and includes features such as Gross Living Area (GLA) reports, 3D videos, and CAD files.

Filemaker 19 (FM19)

Claris International, an Apple Inc. subsidiary, developed Filemaker, a cross-platform database application development program. The MacOS version of the software may be trained to recognize photos using machine learning for use in a wide range of inventory management and other visual recognition applications.

Heidi

Heidi is a SaaS AI-powered medical scribe application that assists clinicians in creating clinical notes during patient visits. Its headquarters are in Cremorne, Victoria. It records consultations, generates notes in the clinician's preferred format, and enables revision and editing.

Invideo
InVideo AI is a San Francisco-based SaaS firm that employs artificial intelligence to translate text into polished movies by automating script development, video editing, background audio, and other production tasks, making it accessible to non-technical users.

Midjourney
San Francisco-based Midjourney is a popular SaaS firm that provides generative artificial intelligence programs for creating images from plain language descriptions, known as prompts, comparable to OpenAI's DALL-E and Stability AI's Stable Diffusion.

Synthesia
Based in London, England, Synthesia is a SaaS firm offering software tools for creating AI-generated video content with personalized avatar presenters without the use of conventional video and editing equipment.

ElevenLabs
AI-driven audio tools startup created in 2022 in New York that creates human-like speech. Their platform enables 32-language text-to-speech, voice cloning, dubbing, speech-to-speech conversion, and text-to-sound effects. ElevenLabs uses contextually aware AI voices to make content globally accessible and improve digital interactions.

Runway Gen-3

An advanced AI tool for video generation, offering high-fidelity, photo-realistic videos from text prompts. It features multimodal training, advanced tools like text-to-video and image-to-video, and fine-grained temporal control for precise key-framing. It excels at generating expressive human characters and provides versatile tools for high-quality video production.

IMPORTANT TO NOTE:

Other organizations and startups will

undoubtedly emerge to challenge their dominant

position as open-source AI improves and

becomes more accessible to a new generation of

creative minds.

Chapter 12

Topics

Your AI Future
Where to Start
Your AI Arsenal
Will AI Replace You
Roll Up Your Sleeves!
Custom GPTs
Moving Forward

Chapter 12

Taking the Next Steps

———————————————————

Whether you like it or not, artificial intelligence and LLMs (Large Language Models) like ChatGPT 4 will change your world. AI has already influenced millions of careers and lives, resulting in some of the world's youngest millionaires. More importantly, it has sparked fear among millions of people around the world about their livelihoods. Are you one of them?

Your AI future

With any technology revolution, many early adopters benefit, while others lag behind and get lost. People will always adapt and accept technology as it becomes more mainstream and levels the playing field. However, on a level playing field, competition can be fierce. If you want to stay ahead of the competition in today's AI age, you must step up your game now before it's too late.

Where to Start

Each week, a host of new AI tools flood the market. Finding the right one to meet your 10X productivity needs can be daunting, like buying a newly issued stock. The same analytical skills and forward-thinking prowess that go into buying a lucrative stock are required to pick the right AI tools for your business and personal needs before the industry-wide shakeout occurs. A similar scenario played out during the personal computer wars in the mid-eighties. When the dust settled, just a handful of companies and businesses were left standing, and fewer are still around today.

Your AI Arsenal

What AI tools should you strap to your ammo belt to come out alive on the other side? Today, AI giants with powerful LLMs like OpenAI's ChatGPT, Microsoft Copilot, Google Gemini, Claude by Anthropic, IBM Watson, and more recently, Grok by Twitter, now known as X, are shaping the future of business and personal productivity. There are a host of other niche AI companies worth keeping an eye on, like Synthesia, which can create virtual avatars from a person's likeness, and Adobe, which has bundled AI automation into its Creative Cloud Suite for content creators, photographers, and graphic designers. Familiarizing yourself with this emerging technology and incorporating it into your workflow will increase your productivity now and prepare you for future advancements in AI technology.

Will AI replace you?

Whether or not artificial intelligence will replace humans in the workforce is one of the biggest worries that millions of people have today. The truth is that AI won't replace people in jobs. People using AI will replace those who don't use it.

Take the legal industry, for example. A custom GPT could allow one paralegal to perform the tasks of five paralegals in a single day, resulting in a 5X increase in productivity. If the same five paralegals used AI, the company's productivity could increase by a factor of 25, enabling them to handle higher caseloads.

Roll up your sleeves!

If you want to get a leg up on the competition with AI technology, you have to get your hands dirty. Sure, you can hire a consultant to pinpoint the pain points in your business and develop an AI solution for your current needs; however, can it stand the test of time as your business needs grow?

One of the first steps is to determine which LLM will best meet the needs of your business. There are a wide variety of LLMs on the market today, each with specific strengths and weaknesses. Configuring the LLM to perform tasks specific to your company's workflow will require creating effective instructional prompts for GPT models. This is a skill that unlocks the real potential of LLMs and machine learning tools. Today, Prompt Engineers, or AI-language model Prompt Engineers, are the backbone or "secret sauce" propelling productivity for businesses and entrepreneurs in the days to come.

Custom GPTs

As a tech evangelist and professional copywriter, I quickly grasped the power that GPTs can unlock for increased productivity. Today, with ChatGPT 4, you can create your own custom GPTs to automate a wide variety of business and personal tasks. The learning curve is lower than you would expect.

For example, I created a GPT called "Pathology Detective" that can list potential physical disorders and symptoms when prompted with lab report findings from blood tests. It has helped me to work more closely with my PCP to monitor and maintain my health.

To help me find coding errors in my FM19 scripts and custom functions, I created "FM19 Expert." This GPT has been a real-time saver and a game changer for me, completely transforming my application development process and workflow. These are just two of the more than dozen custom GPTs. I created it for my personal use and daily workflow. You can take some of them for a spin yourself by visiting www.ai4ceos.com/gpts.

Moving forward

This book is just a brief look at the future of AI and what it can mean for you and your business. We hope it has piqued your interest in this technology and how it can positively transform your business and personal life.
To help you navigate today's AI jungle with the latest news and advances in AI technology that is carefully distilled for easy consumption, please visit our website at www. ai4ceos.com. There, you will find timely blogs and video content to keep you current, well-informed, and highly motivated.

Popular LLMs

The large language models (LLMs) listed here are at the forefront of natural language processing, with cutting-edge features that cater to a wide range of applications such as chatbots and virtual assistants, code generation, automated content creation, and more.

GPT-3

(Generative Pre-trained Transformer 3)

Developer: OpenAI

An early frontier LLM with 175 billion parameters that can generate human-like language, complete sentences, and handle many natural language tasks with accuracy and fluency. It can be used to create chatbots, content, and other applications.

GPT-4

Developer: OpenAI

A larger, more powerful version of GPT-3 that is better at recognizing context, creating more cohesive and contextually relevant. It can write articles, summarize topics, translate languages, and even write stories and poems.

BERT

(Bidirectional Encoder Representations from Transformers)

Developer: Google

A language model that evaluates the entire sentence, both left-to-right and right-to-left, to determine each word's context. This strategy allows BERT to understand the subtle meanings of words based on their context, which improves its performance on a variety of language tasks such as text classification and question answering.

T5
(Text-To-Text Transfer Transformer)
Developer: Google Research

A language model in which all NLP tasks are handled as text-to-text tasks, with text strings as input and output. It can perform a range of NLP tasks, including translation, summarization, and question-answering.

XLNet
Developer: Google Research and Carnegie Mellon University

XLNet is a language model that builds on BERT by employing a permutation-based training strategy to better capture word associations in a sentence. This method produces superior results on natural language processing tasks such as text classification and question answering.

RoBERTa
(A Robustly Optimized BERT Pre-training Approach)
Developer: Facebook AI (Meta AI)
An enhanced variant of BERT trained with greater data and computational capacity, resulting in increased NLP performance. It enhances BERT's pre-training process, resulting in improved text understanding and generation.

ERNIE 3.5
(Enhanced Representation through Knowledge Integration)
Developer: Baidu

ERNIE 3.5 expands on prior models by incorporating more comprehensive pre-training methodologies and larger datasets. The model uses knowledge graphs in its training phase to increase its understanding of complex relationships and contexts within the text. This makes ERNIE particularly effective at tasks that need extensive semantic knowledge, such as question answering, sentiment analysis, and named entity recognition.

MUM
(Multitask Unified Model)
Developer: Google

A multimodal LLM that can interpret and relate information from text, photos, and videos to provide more complete and intelligent responses to user queries. MUM is substantially more powerful than prior models such as BERT, and it aims to improve search engines' ability to handle complicated and multi-step queries.

Megatron-Turing NLG
Developer: NVIDIA and Microsoft

A massive language model with 530 billion parameters, designed to perform a wide range of natural language tasks, from answering questions to generating creative content, with high accuracy and fluency.

PaLM

(Pathways Language Model)

Developer: Google Research

A powerful language model with hundreds of billions of parameters, PaLM can generate, translate, and reason. It's versatile and more efficient, addressing different linguistic tasks with one model instead of multiple models.

LaMDA

(Language Model for Dialogue Applications)

Developer: Google

LaMDA understands and generates contextually relevant responses to make chat discussions feel more natural and fluid. LaMDA can discuss many topics and have natural-sounding conversations, unlike programmed chatbots.

OPT

(Open Pre-trained Transformer)

Developer: Meta AI (Facebook AI)

OPT was created to offer an open-source alternative to huge language models such as GPT-3 with the goal of providing a powerful AI tool for creating and comprehending text while remaining transparent and accessible to the research community.

Jurassic-1

Developer: AI21 Labs

A collection of large-scale language models comparable to GPT-3 that are intended to handle sophisticated text creation and comprehension tasks, with a focus on writing, translating, summarizing, and answering text queries.

GShard

Developer: Google Research

A framework for developing large-scale models with up to a trillion parameters by distributing the task across multiple workstations. This allows for far greater model training than would be possible with a single computer.

GPT-Neo

Developer: EleutherAI

An open-source AI model that generates human-like text from user input. It is comparable to GPT-3 but is free for anybody to use and modify. Popular uses include writing, coding, chatbots, and other creative tasks.

Switch Transformer

Developer: Google Research

An AI model that maximizes efficiency by activating only the most relevant features for a given task. This selective activation reduces processing power usage, making the model faster and more scalable in complex circumstances.

Codex
Developer: OpenAI
A descendant of GPT-3 that has been specifically tailored for translating natural language into programming code comprehension and generation. It powers GitHub Copilot and can help you write, complete, and debug code in multiple programming languages.

Bloom
Developer: BigScience Workshop
A multilingual, open-access language model with 176 billion parameters created by a community of over 1000 researchers to support 46 languages and 13 programming languages, with the goal of increasing transparency in AI research.

Claude
Developer: Anthropic
A collection of large language models created with AI safety and interpretability in mind, with the goal of producing more aligned and controllable results in natural language tasks. The first model was released in March 2023. Claude 3.5 Sonnet, released in June 2024, can identify objects within images.

Demystifying
AI
for Business
Executives

Collins | Taylor II

The following is a curated list of references and resources, including respected academic institutions, reputable websites, and online communities, all offering valuable insights and training on AI technology. These sources offer valuable information for those looking to deepen their understanding and skills in today's constantly advancing field of artificial intelligence.

AI Training Programs

eCornell University

The university offers several online AI Certificate Programs to help equip professionals with practical skills in AI and machine learning. Key programs include Generative AI for Productivity, Designing and Building AI Solutions, AI for Digital Transformation, AI Strategy, and Applied Machine Learning and AI. These programs cover a range of topics, from AI fundamentals and neural networks to business applications and ethical considerations, aiming to help professionals integrate AI into their business strategies and operations effectively.

> *Email: ecornellinfo@cornell.edu*

MIT's Artificial Intelligence: Implications for Business Strategy

A 6-week online curriculum aimed to assist mid to senior-level managers in understanding and leveraging AI technologies like machine learning and robots in their organizations. It offers a self-paced learning experience with interactive videos, quizzes, assignments, and discussion forums guided by a Success Adviser. Participants examine AI use cases, ethical issues, data strategy, and organizational structure for incorporating AI into their business plans and earn a certificate of completion from the MIT Sloan School of Management.

> *Email: mitsloan@getsmarter.com*

Stanford's AI for Business Professionals

This online program is tailored for non-technical professionals seeking to understand and apply AI in their businesses. It offers foundational knowledge of AI and strategies for its integration into operations. Key courses include "Generative AI: Technology, Business, and Society," focusing on AI fundamentals and societal impacts; "Mastering Generative AI for Product Innovation," a guide for product leaders; and "Building an AI-Enabled Organization," covering AI strategy, data architecture, and business process integration.

> *Email: enterpriseeducation.stanford.edu.*

UT Austin McCombs School of Business

The Artificial Intelligence and Machine Learning: Business Applications is a 7-month online certificate course designed to arm applicants with job-ready AI and ML skills. The program offers interactive mentor-led sessions, covers AI and ML fundamentals like neural networks and NLP, and features 8 hands-on projects. The program also features personalized career support and optional on-campus immersion. Graduates earn a professional certificate from UT Austin.

> *Email: aiml.utaustin@mygreatlearning.com*

Websites

AI4CEOS.COM

The official website of this book is where you can find updated content on the latest advances in AI technology, chatbots, case studies, and more.

AIhub.org

An online platform operated by the Association for the Understanding of Artificial Intelligence, a non-profit organization. Its goal is to deliver free, high-quality material direct from AI professionals to both the AI community and curious technology fans via main-stream media coverage of their articles.

Futuretools.io

A comprehensive platform that aggregates AI tools to help users locate what they need. It offers productivity, marketing, music, generative art, video editing, and more. View ratings, features, and prices of tools by category, filter, or search. Users can identify and use AI technologies on the site to improve their skills and productivity.

Gptstore.ai

The GPT Store is where users can discover and utilize various GPT-based tools. It features a wide range of AI-driven applications, from image generation and creative writing to research assistance and video production. Users can explore different GPTs, sorted by categories like top-rated, virtual assistants, and creative tools, to find the ones that best suit their needs. The site is community-driven, allowing creators to share and promote their GPTs.

Myaiadvantage.com

A platform designed to help users improve interactions with AI technologies such as ChatGPT. It offers tools such as YouTube videos, public speaking engagements, newsletters, and a network to assist people in using AI for personal and professional development. The platform focuses on practical AI applications to help increase productivity and entrepreneurial initiatives.

Curiousrefuge.com

An online platform for AI filmmaking and storytelling. It provides training and materials to help creators use AI to make films, advertise, and engage in other creative endeavours. The platform's goal is to provide accessible training and resources for AI cinema fans and pros while also pioneering a new approach to using AI in visual storytelling.

Sakana.ai/ai-scientist

AI Scientist by Sakana is a revolutionary system designed to automate scientific discoveries completely. It can independently brainstorm, code, conduct experiments, and write a scientific paper. This method, developed with Oxford and British Columbia, promises to democratize and accelerate scientific development by making research more efficient and cost-effective, with each research paper costing $15.

Online AI Communities

AI Society: aisociety.in

A group dedicated to increasing global access to artificial intelligence. They provide online and offline courses, boot camps, and certification programs to help people, entrepreneurs, organizations, and college students take their initial steps into AI and master advanced topics.

AI Village: aivillage.org

A group of hackers and data scientists are actively working to spread awareness about how AI impacts security and privacy. They aim to bring in diverse voices and build a community focused on making AI safer. They play an active role at DEF CON, the world's biggest hacking conference.

Kaggle AI Community: Kaggle.com

With over 15 million members, Kaggle offers a platform for refining and showcasing machine-learning skills through competitions and projects.

AI Stack Exchange: ai.stackexchange.com

A Q&A community where you can ask questions and get answers from AI experts.

Hugging Face: huggingface.co

A community where AI enthusiasts and professionals collaborate, share ideas and explore AI capabilities. Popularly known as the "GitHub of machine learning," where users can test machine learning models from its vast library of over 28,000 models.

OpenAI Forum: forum.openai.com

A unique program that brings topic experts and students together to discuss and plan AI's future. In-person technical discussion meet-ups, OpenAI dinner mixers, educational webinars, expert roundtable conversations, and lots of networking and idea-sharing opportunities are offered at the Forum.

Towards AI: towardsai.net/community

Launched in 2019, Towards AI is a leading educational resource and community platform for AI enthusiasts, practitioners, and students for sharing information, educational content, and research on AI, with over 2,000 authors and hundreds of thousands of followers. It covers a wide range of topics, including machine learning, deep learning, NLP, computer vision, and more.

Glossary of Key Terms

Adaptive Learning	An educational system that employs artificial intelligence to personalize learning experiences by tailoring curriculum and assessments to specific students' needs and performance levels
Adversarial Attack	A technique for fooling machine learning models by introducing modest, purposeful modifications to input data, which is frequently used to assess model resilience.
Agent	An AI software model that can execute specific tasks autonomously on behalf of a user with or without human interaction.
Algorithm	A set of rules or instructions that an AI model follows to learn from data and make judgments.
Algorithmic Bias	The existence of systematic and unfair discrimination in the outcomes produced by AI systems, generally due to biased training data or model design.
Artificial General Intelligence (AGI)	A theorized form of artificial intelligence capable of performing any cognitive function that a person can.
Artificial Intelligence (AI)	Computer systems capable of simulating human intelligence processes.
Artificial Narrow Intelligence (ANI)	AI systems that specialize in a single task or a limited set of tasks.

Artificial Neural Network (ANN)	A computational model based on biological neural network structure that is employed in a variety of machine learning applications.
Artificial Superintelligence (ASI)	A fictional artificial intelligence that outperforms humans in all areas, including creativity and social abilities.
AUC (Area Under the Curve)	A numerical metric that determines a binary classification model's overall ability to decern between positive and negative classes. It is measured by the area below the receiver operating characteristic (ROC) curve. An AUC of 1.0 represents perfect classification, whereas an AUC of 5 indicates performance equivalent to random guessing.
Augmented Reality (AR)	A technique that superimposes digital information, like as images or data, on the real world to improve the user's impression of reality.
Autoencoder	A type of neural network used for unsupervised learning that learns efficient representations (encoding) of data, typically for dimensionality reduction or noise reduction.
Automated Machine Learning (AutoML)	A system that simplifies and accelerates the process of developing, selecting, and optimizing machine learning models for real-world applications.
Automation	The application of technology to accomplish activities with little or no human participation resulting in enhanced efficiency and production.
Autonomous Vehicle	A self-driving car or other vehicle that uses AI and sensor systems to navigate and operate without human input.
Backpropagation	A learning algorithm used by neural networks which adjusts for mistakes, making the network better at predicting the right answers.

Bagging	A learning strategy that involves training many models on distinct random sections of the data and then combining their results to produce a more accurate and trustworthy outcome.
Batch Normalization	A strategy for making neural networks train quicker and more precisely by modifying and scaling data as it travels through the network.
Bayesian Optimization	A clever method for determining the best model parameters by leveraging previous findings to guess and test the most promising possibilities next.
Bias	Bias in machine learning occurs when a model produces errors because it oversimplifies the problem or because of training decisions, such as the type of model used or the quality of the training data. This can cause the model to overlook important patterns.
Bias-Variance Trade-off	A machine learning design method that seeks to strike the optimal balance between making a model too simple (which misses patterns) and too complex (which overreacts to minor details) to create the most accurate predictions.
Big Data	Refers to massive, complicated datasets that are too large for typical data processing methods to manage. More advanced technologies are usually required to store, analyze, and derive insights from the data, which frequently comes from various sources at high speeds.
Big Data Analytics	The practice of studying huge and complicated datasets to discover useful patterns and insights that can help businesses make better decisions.
Boosting	An ensemble learning strategy that improves accuracy by combining predictions from multiple weak learners.
Bot	A software model that is designed to execute automatic functions like as answering questions, sending messages, and surfing the internet without the need for human assistance.

Capsule Network	A type of neural network that better understands the relationships between visual elements of an image, making it more accurate at recognizing objects.
Chat-bot	A software model that can speak with humans, typically via the Internet, using natural language processing to mimic human speech.
Classification	A machine learning method in which a model is trained to categorize data into several groups or labels.
Clustering	A machine learning technique in which data points are grouped according to their similarities, resulting in items in the same group (or cluster) being more similar than those in other groups.
Cognitive Computing	A system that uses artificial intelligence to process information, understand natural language, and learn from data in order to assist in the solution of difficult problems and decision-making.
Cognitive Load	The amount of mental energy and resources needed to comprehend information, acquire new skills, or solve problems. High cognitive loads are more difficult to learn and remember knowledge, whereas low cognitive loads are easier to manage and comprehend.
Computer Vision	A table that displays how well a classification model performs by comparing its predictions to the actual results, emphasizing where it gets things right and where it makes mistakes.
Confusion Matrix	Refers to data that does not follow a predefined model or format, such as free-form text, images, videos, or audio, which the AI must interpret or generate content from without relying on a rigid structure.
Conversational AI	Technology that enables computers to communicate with humans in a natural manner, such as chatbots or virtual assistants that can understand and respond to what you say or type.

Convolutional Neural Network (CNN)	A type of artificial neural network that is particularly good at recognizing patterns in images, such as detecting objects or faces, by analyzing the image in little sections at a time.
Cross-Validation	A technique for evaluating the performance of a machine learning model that divides the data into various parts trains the model on some of them and tests it on the rest.
Data Augmentation	A machine learning approach for increasing the amount and variety of a training dataset by modifying existing data, such as flipping, rotating, or zooming in on photographs, in order to improve model performance and generalization.
Data Governance	The management system that ensures data availability, integrity, security, and appropriate use inside a company.
Data Mining	The process of identifying patterns, trends, and relevant information in massive datasets using techniques such as statistical analysis, machine learning, and database systems.
Data Pipeline	A set of operations and procedures that transfer data from one system to another, frequently including data collection, cleansing, transformation, and loading in order to be used for analysis or other reasons.
Data Science	A discipline that integrates statistics, computer science, and domain expertise to extract valuable insights and knowledge from data, frequently employing techniques like machine learning, data mining, and visualization.
Decision Tree	A model used in machine learning and data analysis that splits data into branches depending on specific conditions, similar to a flowchart, until it reaches a final judgment or prediction.

Deep Learning (DL)

A form of machine learning that employs highly complicated networks with numerous layers to learn from massive quantities of data, commonly used for image recognition and speech comprehension.

Digital Assistant

Voice-activated digital assistants (such as Siri, Alexa, or Google Assistant) are software programs that can interpret and respond to user-inputted text or speech commands.

Digital Twin

A virtual version of a physical object, system, or process that is used to replicate, analyze, and monitor its real-world counterpart in real-time, thereby improving performance and predicting problems.

Dimensionality Reduction

A data analysis and machine learning technique that reduces the amount of input variables (features) in a dataset while maintaining as much critical information as possible, allowing the data to be visualized and processed more easily.

Dropout

A regularization strategy used in neural networks to minimize overfitting by randomly "dropping out" (disabling) a fraction of neurons during training, causing the network to learn more robust and broad characteristics.

Dropout Layer

A neural network layer that switches off some neurons at random during training to reduce overfitting and improve model robustness.

Edge AI

AI that runs locally on local devices such as smartphones or cameras, enabling fast results without the need to send data to a central server.

Edge Computing

When computing processes data directly where it's created, such as on local devices, rather than sending it to a remote server, making things faster and less reliant on the internet.

Ensemble Learning	A machine learning strategy in which numerous models are integrated to increase overall performance, resulting in more accurate predictions than a single model alone.
Ethical AI	Ethical AI involves developing and deploying AI systems in a fair, transparent, and responsible manner that avoids harm, bias, and discrimination while also respecting privacy and human rights.
Expert System	A computer software that simulates a human expert's decision-making skill by applying a set of rules and knowledge to solve specific problems in fields such as medicine, economics, and engineering.
Explainability	The ability to grasp and explain how an AI model makes its judgments, making its processes transparent and interpretable for people, and assuring trust and responsibility.
Explainable AI (XAI)	Artificial intelligence systems that make their decision-making processes transparent and understandable allowing consumers to better understand how and why a particular choice was made, hence boosting trust and accountability.
F1 Score	A method used to evaluate a model's performance by comparing the number of positive predictions against inaccurate predictions, where the final result is represented as a percentage score.
Feature Engineering	The process of preparing and developing meaningful data inputs (features) from raw data to assist a machine learning model in making more accurate predictions.
Federated Learning	A method for training machine learning models on data distributed across multiple devices without moving the data off the devices, hence keeping the data private and secure.

Fuzzy Logic	A method of thinking that deals with uncertainty by letting something to be partly true or partly false rather than totally true or false, which is helpful in situations where things are not clear.
Generative Adversarial Network (GAN)	The system consists of a competitive setup between two neural networks, with one network generating counterfeit data and the other attempting to recognize the counterfeit. Over time, the system develops its ability to generate real data, such as highly convincing pictures or audio.
Gradient Clipping	A mechanism used during training to keep changes in the model's weights from becoming too large, hence stabilizing the learning process and preventing the model from crashing.
Gradient Descent	A method used in machine learning to improve a model by slowly adjusting its settings to reduce errors until it reaches an optimum outcome.
Hierarchical Clustering	A method of grouping comparable data points together to form clusters arranged in a tree-like structure, either by merging smaller groups into larger ones or by breaking a large group into smaller ones.
Human-in-the-Loop (HITL)	When humans interact with machines to train, refine, or make judgments in a system, resulting in greater accuracy and managing circumstances that require human judgment.
Hyperparameters	Settings or configurations in a machine learning model that are specified before training, such as the learning rate or the number of layers in a neural network, which impact how the model learns and performs.
Internet of Things (IoT)	A network of physical devices connected to the internet and capable of collecting, sharing, and acting on data.

Glossary of Key Terms

K-Means Clustering	A method for grouping data into a given number of clusters (k), with each data point belonging to the cluster with the nearest mean, which aids in organizing comparable objects.
K-Nearest Neighbors (KNN)	A network of linked data points or concepts that represent real-world entities and their relationships, allowing for more efficient organization and retrieval of information.
Logistic Regression	A machine learning algorithm that predicts whether something falls into one of two categories (such as "yes" or "no") based on input data.
Long Short-Term Memory (LSTM)	A neural network that can recall key information over time, making it ideal for tasks such as voice or text comprehension.
Loss Function	Measures how inaccurate a model's predictions are, allowing the model to improve by minimizing this error.
Machine Learning (ML)	Training technology where computers learn from data and make predictions or judgments without being directly programmed.
Machine Vision	Computer vision technology that enables computers to perceive and interpret images or videos, allowing them to perform tasks such as item identification, text reading, and robot guidance.
Natural Language Processing (NLP)	Technology that allows computers to recognize, interpret, and reply to human language, whether written or spoken.

Natural Language Understanding (NLU)	A branch of NLP concerned with teaching computers to understand the meaning and intent of human language, allowing them to comprehend context and nuances.
Neural Architecture Search (NAS)	An automated deep learning approach that finds the best neural network design architecture for a specific task without manual refinements. The above minimizes the time and expertise needed to develop effective neural networks, making them more accessible for complex applications.
Neural Machine Translation (NMT)	A system that uses artificial intelligence to translate text between languages by comprehending whole sentences, resulting in more accurate and natural translations.
Neural Network	A computer system modeled after the human brain, consisting of layers of interconnected nodes (neurons) that collaborate to spot patterns, make decisions, or solve problems by learning from data.
Neural Turing Machine (NTM)	A neural network that can learn and use memory, much like a computer, to perform tasks that require data storage and recall.
Optical Character Recognition (OCR)	A technology that translates text pictures, such as scanned documents or photos, into digital text that can be edited and searched.
Overfitting	Overfitting is when a model performs well on the training data but does poorly on new data because it focuses too much on details that don't matter.
Personalization	The process of customizing content, recommendations, and experiences for individual users based on their preferences, behaviours, and data.

Glossary of Key Terms

Precision	The ratio of actual positive predictions to the model's total number of positive predictions indicates positive prediction accuracy.
Predictive Analytics	The application of statistical algorithms and machine learning approaches to predict future outcomes based on past data.
Quantum Computing	The principles that allow quantum computers to execute computations in ways that ordinary computers cannot, such as the use of qubits, which can represent both 0 and 1 at the same time, allowing for faster processing of complicated problems.
Random Forest	A machine learning technique that generates a large number of decision trees and combines their results to make more accurate predictions.
Recall	The measure of true positive predictions to total positive instances in the data which indicates the model's ability to correctly identify positive instances.
Recommendation System	An AI-powered system that recommends items, services, or content to users based on their interests, behavior, and previous interactions; often used by platforms such as Netflix and Amazon to help people find what they would like.
Recurrent Neural Network (RNN)	A neural network that can recall previous information, making it ideal for sequence-based tasks such as predicting the next word in a sentence.
Regression	A machine learning statistical technique that predicts a continuous numerical value from input data.

Reinforcement Learning (RL)	A training method where a system learns by doing tasks, receiving feedback, and refining its behaviors in order to earn more positive rewards.
Robotic Process Automation (RPA)	A system that uses software bots to automate repetitive processes, such as data entry or transaction processing, speeding up work and eliminating errors.
ROC Curve (Receiver Operating Characteristic Curve)	A chart that indicates how successful a model is at distinguishing between two classes, with better models having curves closer to the graph's top-left corner. See also Area Under the Curve (AUC).
Semi-Supervised Learning	A learning strategy that trains a model with a small quantity of labeled data and a large amount of unlabeled data, allowing the model to learn more efficiently when labeled data is limited.
Sentiment Analysis	A system that uses artificial intelligence to assess the emotional tone of a piece of text, such as whether it is favorable, negative, or neutral; it is frequently used to analyze opinions in reviews, social media, or customer feedback.
Smart City	A city that employs technology to increase service efficiency, improve people' lives, and reduce environmental effect, such as smart traffic signals, real-time public transit alerts, and energy-saving utilities.
Smart Contract	A digital agreement written in code that automatically enforces its terms when certain conditions are satisfied, eliminating the need for a mediator.
Smart Home	A home outfitted with IoT devices that use AI to automate and control numerous services, such as lighting, heating, and security, frequently using voice requests or smartphone apps.

Glossary of Key Terms

Speech Recognition	The ability of a machine or software to recognize and analyze human voice, resulting in text or commands.
Supervised Learning	A training strategy in which a model learns from data with correct responses in order to make accurate predictions or judgments.
Support Vector Machine (SVM)	A machine learning method for determining the optimal line or border to separate various groups of data in tasks like classification.
Support Vector Regression (SVR)	A type of machine learning technique based on Support Vector Machines (SVM) but designed for regression problems in which the goal is to predict a continuous value rather than classify data into categories.
Swarm Intelligence	The collective behavior of decentralized, individual agents, such as robots or drones that, collaborate to solve complicated problems comparable to the social behavior of ants, bees, and fish.
T-SNE (t-Distributed Stochastic Neighbor Embedding)	A method for converting complex dimensional data into 2D or 3D so that it can be viewed, making it simpler to discover patterns and groupings by clustering similar data points together.
Test Data	Data used to assess a machine learning model's performance following training.
Tokenization	Natural language processing technique, which involves breaking down text into smaller parts, like words or phrases, for easier analysis.

Training Data	A dataset used to train a machine learning model by displaying examples with known outcomes, allowing it to make accurate predictions on new data.
Transfer Learning	A machine learning strategy in which a model created for one task is utilized as the foundation for a model on another task.
Turing Test	A testing procedure devised by Alan Turing to determine whether a machine demonstrates intelligent behavior indistinguishable from that of a person.
Underfitting	A scenario in which a machine learning model fails to detect the underlying pattern in the data, resulting in poor performance on both training and new data.
Unsupervised Learning	A type of machine learning in which the algorithm learns from data without explicit instructions, recognizing patterns and relationships on its own.
Variational Autoencoder (VAE)	A generative model that learns to compress data into a smaller form and then replicate it with minor alterations, resulting in new data that is similar but not identical to the original.
Virtual Reality (VR)	A technology that generates a virtual, immersive environment in which users can engage, generally via a headset that records their motions and displays 3D graphics, giving the impression that they are in another world.
Voice Assistant	A software program that recognizes and answers to spoken commands, assisting users in things such as setting reminders, playing music, and answering inquiries; commonly found in smartphones and Apple's Siri, Amazon's Alexa, or Google Assistant.